EXTREME BODY PARTS

BLACKBIRCH PRESS

An imprint of Thomson Gale, a part of The Thomson Corporation

THOMSON

★

™

GALE

Detroit • New York • San Francisco • San Diego • New Haven, Conn. • Waterville, Maine • London • Munich

For more information, contact
Blackbirch Press
27500 Drake Rd.
Farmington Hills, MI 48331-3535
Or you can visit our Internet site at http://www.gale.com

Photo credits: Cover: top left, top center, middle right © Discovery Communications, Inc.; top right © Digital Vision; middle left, bottom left, bottom right Corel Corporation; all pages © Discovery Communications, Inc. except for pages 1, 28, 36 Corel Corporation; page 4 © Royalty-Free/CORBIS; page 8 © Simon Foale/Lonely Planet Images; page 16 © Nigel J. Dennis; Gallo Images/CORBIS; page 20 © Photos.com; page 24 © Joe McDonald/CORBIS; page 29 © Tom Brakefield/CORBIS; page 32 © Reuters/CORBIS; page 40 © PhotoDisc

LIBRARY OF CONGRESS CATALOGING-IN-PUBLICATION DATA

Body parts / John Woodward, book editor.
    p. cm. — (Planet's most extreme)
    Includes bibliographical references and index.
    ISBN 1-4103-0395-0 (hardcover : alk. paper) ISBN 1-4103-0437-X (paper cover : alk. paper)
    1. Anatomy, Comparative—Juvenile literature. I. Woodward, John, 1958– . II. Series.

    QL806.5.B63 2005
    571.3'1—dc22
                                                                                          2004017183

Printed in the United States of America
10 9 8 7 6 5 4 3 2 1

If you think you have a beautiful body, just wait till you see the unique physiques found in the natural world. We're counting down the top ten most extreme body parts in the animal kingdom and seeing how we measure up when body parts are taken to The Most Extreme.

10

The Fennec Fox

Our countdown of extreme body parts starts in the scorching heat of the Sahara desert. We're on the trail of an animal that really keeps its ear close to the ground. It's the fennec fox.

The enormous ears of the fennec fox help it to hunt. If you had ears like this fox, they'd hang down to your chest!

Smaller than a house cat, this is the smallest fox in the world, but it has a serious set of ears. The fennec fox is number ten in the countdown because of its enormous ears. They act like furry satellite dishes, funneling the smallest sounds into the eardrum so that even on the darkest nights, the fox can target the tiniest creatures creeping over the sand.

Imagine if you had ears like a fennec fox. You'd have to have ears that measured one-third the length of your body! If we were like the fennec fox, we'd have ears nearly twice the size of our feet!

5

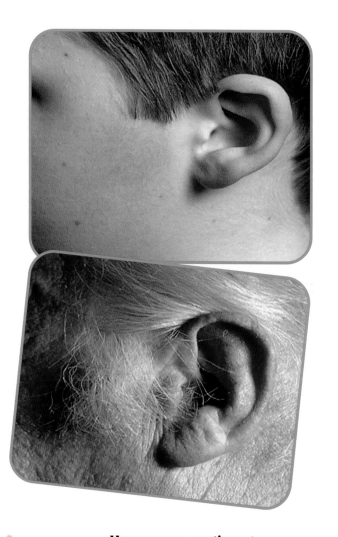

Actually, our ears keep growing with age. They grow about 1/800th of an inch every year. We'd have ears like a fennec fox if we lived for about 16,000 years!

Not all ears are equal. Everybody's ears are different, even on your own head! And some researchers believe this can be a sign of how our brain works! A group of Russian scientists measured the differences in length of people's left and right ears. They concluded that if your left ear is slightly longer than your right, you might have analytical abilities associated with the sciences. If your right ear is longer, you might have a tendency toward the creative arts. The theory is that the difference in ear length reflects which side of the brain is dominant.

Human ears continue to grow as we get older, adding a fraction of an inch in length every year.

You might say the fennec fox keeps cool in the Sahara desert with "ear conditioning." Its ears carry heat away from the body.

The fennec fox's big ears have another function in the desert, as biological educator Paul Hahn explains:

These guys, despite their small size, have the largest ear to body ratio of any canid—canids are animals like dogs, wolves and foxes. They have adaptations in their ears that allow them to dissipate heat, and we like to joke around and say that they have ear conditioning!

These big ears work just like the radiator in a car. They're packed full of blood vessels close to the surface. The blood carries heat away from the engine and radiates it out through the massive ears.

9

The **Platypus**

Paddling into number nine on the countdown is the platypus. While it's no monster, it is such an extreme collection of body parts that it could have come straight from Frankenstein's laboratory. All you need are the four webbed feet of an otter, one paddlelike tail of a beaver, and the beak of a duck.

The platypus is made up of an odd grouping of body parts, including webbed feet, a beaverlike tail, and a duck bill.

Frankenstein would be proud of this collection of spare parts. It's easy to see how it uses the webbed feet as propellers and the tail as a rudder. But that's no ordinary duck's bill. It's actually packed with hundreds of receptors that detect the tiny electric currents produced by bugs moving in the water. No wonder early naturalists couldn't believe their eyes.

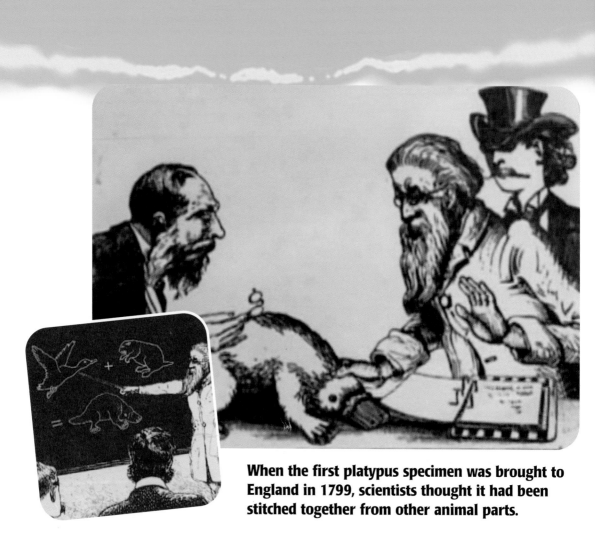

When the first platypus specimen was brought to England in 1799, scientists thought it had been stitched together from other animal parts.

In 1799, explorers returned to England with the first platypus specimen. It was greeted with such skepticism that naturalists dissected the pelt expecting to find stitches attaching the bill to the skin! They thought the platypus was a hoax.

In 1912, an amateur archaeologist working in the Piltdown quarry in Sussex, England, discovered what seemed like ancient human remains. For over 40 years the "Piltdown man" was celebrated as the "missing link" that combined the features of both human and ape. And it was British!

It wasn't until the 1950s that more sophisticated archaeological dating techniques exposed the Piltdown man as an impostor! It was revealed that the skull was human but only 500 years old. The jaw belonged to an orangutan, and the teeth came from monkeys and hippos! Even today no one really knows who planted the Piltdown plot.

Nobody doubts the platypus, though. This extreme collection of body parts has a fossil record going back more than 100 million years!

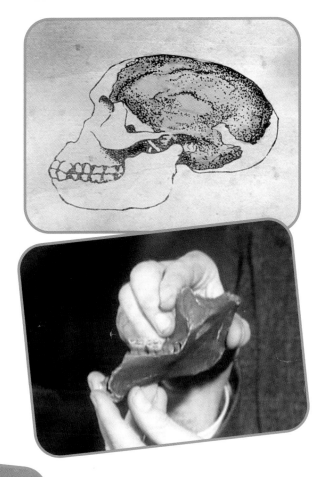

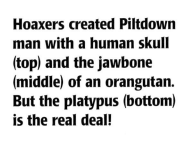

Hoaxers created Piltdown man with a human skull (top) and the jawbone (middle) of an orangutan. But the platypus (bottom) is the real deal!

11

8

The Babirusa

In the tropical forests of Indonesia, our next contender in the countdown of extreme body parts loves rain. Lots of rain means lots of mud, and that's great for the secretive animal that's number eight in the countdown.

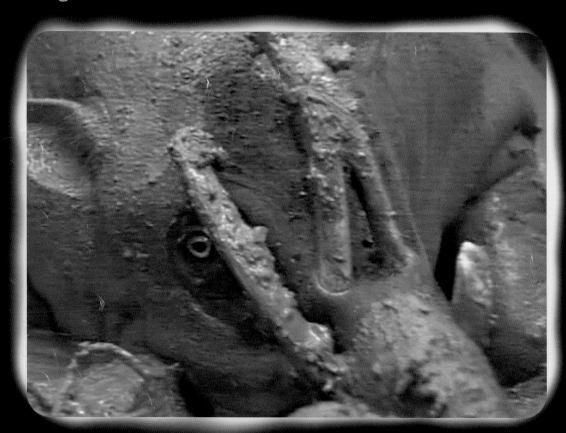

This extraordinary pig is a dentist's worst nightmare. It has two teeth growing straight through the roof of its mouth! These extreme tusks are actually crazy canine teeth! The babirusa is number eight in the countdown because unlike our teeth, babirusa tusks keep on growing. Sometimes they curl around so far that they can grow right into their own skull! But the upper canines are mostly for show. It's the lower tusks that are used as weapons, and are sharpened regularly by rubbing against trees.

It must be really uncomfortable having a mouth full of such extreme dentition. How would you like to have foot-long canine teeth growing out the roof of your mouth? An orthodontist could make a lot of money correcting that overbite.

The babirusa's tusks (top) are actually teeth that never stop growing. Think of all the trips to the dentist you'd make if you had such long teeth (bottom).

Your canine teeth (inset) are the largest in your mouth, but they are tiny compared to the canines of primates like this baboon.

While our canines may not be as impressive as a babirusa's, they're still the biggest teeth in our mouth. But compared to those of the rest of our primate cousins, our canine teeth are very small indeed.

That's because male apes use big sharp canines for aggressive displays and as weapons to defend themselves.

When humans expose their teeth, sometimes it can be a sign of happiness. But the human smile is also closely related to the primate "fear face." A public baring of the teeth is the chimp's way of declaring it's scared. So the next time you're on a roller coaster, take a look to see if the humans are happy or if they're wearing their primate "fear face!"

For the babirusa—exposing your canines makes the ultimate "fear face." One look at that extreme dentition is usually enough to scare away any rival males!

Although humans smile when they are happy, the chimpanzee bares its teeth to show fear.

The Aye-Aye

In Madagascar, the next contender in the countdown of extreme body parts really has its finger on the pulse of the forest. Meet the aye-aye. It may look like a cross between an opossum and a rat—but it's actually a very peculiar primate.

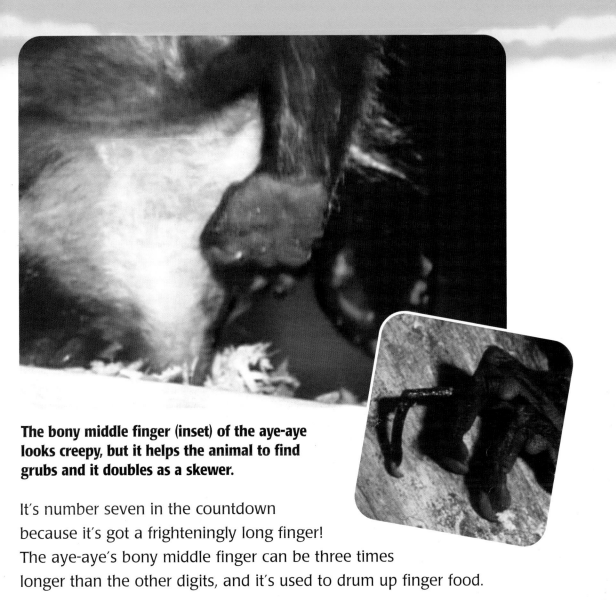

The bony middle finger (inset) of the aye-aye looks creepy, but it helps the animal to find grubs and it doubles as a skewer.

It's number seven in the countdown because it's got a frighteningly long finger! The aye-aye's bony middle finger can be three times longer than the other digits, and it's used to drum up finger food.

Scientists have discovered that the aye-aye finds food simply by tapping its finger. Those big floppy ears are listening for the difference between the solid wood and the hollow tunnels filled with juicy grubs. Having accurately located the grub with its fingertip Morse code, it's a simple matter to rip apart the wood, and use the versatile drumstick to extract a meal.

More extreme than the aye-aye's bony finger are Ruth Ward's nine-inch fingernails.

The aye-aye may be blessed with an extremely useful digit, but in Britain, there's one woman whose dramatic digits are a bit of a curse. Meet Ruth Ward. She has nine-inch-long fingernails. Living with these extreme nails isn't easy. She explains:

> I have to park my car in gear. I can't use a handbrake. I can't do jewelry up. I can't do a button up.

It took Ward three long years to grow her fingernails. So why would anyone want nails so long that to paint them all requires three bottles of nail polish? Ward continues:

I just got carried away with the idea to see them grow and they grew.

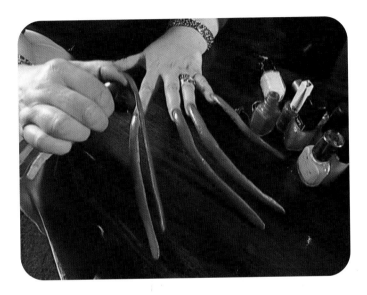

Even Ruth Ward would be impressed with the aye-aye's incredible finger. It's lucky she hasn't seen the next contender, because then she might have been tempted to grow her toenails, too.

It took three years for Ruth Ward to grow her fingernails. Even she would envy the aye-aye's amazing middle finger!

6

The Jacana

In parts of South America, there can be a distinct shortage of dry land. So it's handy if you have an extreme body part that let's you walk on water. And that's why the jacana tiptoes into number six in the countdown. Its feet are so big that it can run over floating vegetation. No wonder it's also called the "lily trotter."

It sure would be hard to find shoes that fit if we had gigantic feet like the jacana!

Imagine if we had feet like a jacana. Unlike the jacana, we'll never be able to trot across lilies because our feet are just too small. Your average foot is only 15 percent of the total length of your body. But if we had feet like the jacana, they'd be well over three feet long! That's an incredible 60 percent of your body length! Things with feet that size can be really scary!

Could this be Bigfoot caught on tape? No real proof of Bigfoot's existence has been found yet.

Bigfoot is one of the great American legends. Hundreds have searched the forests of North America for evidence of this giant ape-man. But apart from sightings of very large footprints, undeniable proof of its existence has yet to be found. But Bigfoot may soon have lots of company, for it's living in a nation of increasingly big feet!

The average American woman's foot was a dainty size 3½ back in 1900. But over the years, thanks to an improved diet, American women grew bigger and taller—and their feet kept pace. Now the average woman's foot is a stylish size 8!

Of course, we've got a long way to grow if we want to catch up to the jacana. This is one bird that really does have big shoes to fill!

In one century, the average size of the American woman's foot grew from a tiny size 3½ in 1900 to a size 8 today.

1900 SIZE 3½

2000 SIZE 8

The **Fiddler Crab**

The next contender in our countdown of extreme body parts never has to work out. And yet not even the strongest muscle man is a match for the most amazing arms in the natural world.

The male fiddler crab would be great at hitchhiking. He uses his enormous claw to attract females and to challenge other males.

You find them a long way from Muscle Beach. They're more at home in the sticky ooze of tropical mudflats. Crawling into number five in the countdown is the fiddler crab.

This male has one claw that's massively enlarged. A big claw is a big help when it comes to finding a mate. Different species of fiddler crabs have different colored claws, and they spend a lot of time waving them around. This is both an invitation for the females to mate and a challenge to rival males.

It may look like these fiddler crabs are out to hurt each other, but they really only arm wrestle.

Fiddler crab fighting is extreme arm wrestling. In the human world, owners of the biggest biceps fight it out on stage, but the only thing these males strike is a pose! Body builders can bulk up their biceps until they're twice the size of the average male's arm. In fact, the best musclemen have arms that are bigger than the average guy's thigh!

But no matter how much iron you pumped, you'd still never bulk up enough to match the fiddler crab. That's because our arms usually make up about 6 percent of our body weight. If we were fiddler crabs, we'd look very different. Our arm would be ten times bigger! No wonder the fiddler crab has trouble carrying around that huge claw. It's 65 percent of its body weight!

When the tide goes out on the mudflat, the race to feed is on. And that's when the male's big claw can be a burden. It's so big that he can't use it to feed. He has to pick tasty treats off the mudflat using only one arm, so he spends twice as long feeding as the female.

Shake hands with a man whose arm is 65 percent of his body weight (top)! With two small claws, the female fiddler crab has no trouble eating (bottom).

27

4 The **Anteater**

In the grasslands of Brazil, wherever there are termite mounds you can be sure to find our next contender. For termites are one of the favorite foods of the giant anteater.

The giant anteater's sticky tongue can grow to more than two feet. It's perfect for finding tasty ants and termites.

This bizarre collection of seemingly mismatched body parts is built for one thing—to seek and destroy ants and termites. That long nose comes with an extreme sense of smell, and its forearms and claws are so powerful that the giant anteater can rip open a termite mound or anthill with a single blow of its paw.

The anteater is number four in the countdown because of what happens once it has found a nest. The tongue of the anteater can be over two feet long! It's covered in sticky mucus and can be poked down a termite tunnel 150 times a minute! This extreme body part puts our tiny tongue to shame.

At nearly three inches, Umar Alvi's tongue is one of the world's longest. But it's still no match for the anteater's tongue.

The average human tongue measures about an inch from lip to tip. But this tongue belongs to the human equivalent of the giant anteater. Umar Alvi from London is the proud holder of one of the longest tongues in the world—stretching over 2½ inches from tip to lip. But no matter how you measure it, his tongue is still no match for the giant anteater's.

Licking up to 30,000 termites each day gives the giant anteater an unexpected advantage in the struggle for survival. It has developed a tolerance to the normally toxic formic acid found in its diet of ants and termites. This poison accumulates in its body, which means that few predators could stomach an adult anteater. But baby anteaters are still tasty because they haven't eaten enough toxic termites. That's why they hide on top of mom's back until their amazing tongue has helped them to lick all opposition.

Baby anteaters are very tasty to predators. That's why the babies protect themselves by riding on mom's back.

The Giant Squid

So far we've seen anteaters getting tongue-tied, crabs lending a helping hand, and jacanas putting their foot in it. But what is the next contender? Meet number three in the countdown—the giant squid. Its monstrous body parts have baffled scientists for centuries.

Most of what we know of these enormous animals comes from specimens that were found dead on the shore. That's because a living giant squid has never been filmed, and so we have to imagine what they're like in the wild.

We know that like all squid they can move slowly by gently undulating their fins, or rapidly by jet propulsion. We also know that they hunt using their tentacles and a beak like a parrot's that crushes their food.

But what makes this squid number three in the countdown is the sheer size of its body parts. It's not hard to see why these animals were once thought to be monsters. Imagine swimming next to an animal that's the same size as a truck and semitrailer. That's nearly 60 feet of squid! So it's no surprise that this giant has the largest eyes on the planet. The squid's eyes have a 15-inch diameter—making them about the size of large beach balls.

57
FEET

Although the giant squid is as big as a truck and trailer, nobody has ever seen one alive. Scientists only study dead specimens.

Our puny eyeball is only one inch across, which is smaller than a golf ball! Despite their relatively small size, human eyes are packed with some sophisticated technology. Lining the back of the eye is the retina, which lets us see the world as a pattern of tiny dots—just like a newspaper photograph. That's because the retina is made up of specialized detector cells—125 million of them.

The giant squid's eyeballs are the world's largest. They're as big as beach balls!

15 INCHES

There are more than a billion detector cells lining the back of the giant squid's eyeball—and it needs every last one of them in the darkness of the deep ocean. By watching this arrowhead squid in action, we can get an idea of how the giant squid hunts. Once its massive eyes have locked onto the target, it fires two clasper tentacles to pull in its prey.

Scientists believe that giant squid live on a diet of fish. But perhaps it's lucky that we've never delved too far into the deep. After all, would you like to catch the eyes of this gigantic predator?

Just like the arrowhead squid shown here, the giant squid probably hunts by firing clasper tentacles that pull in dinner to its mouth.

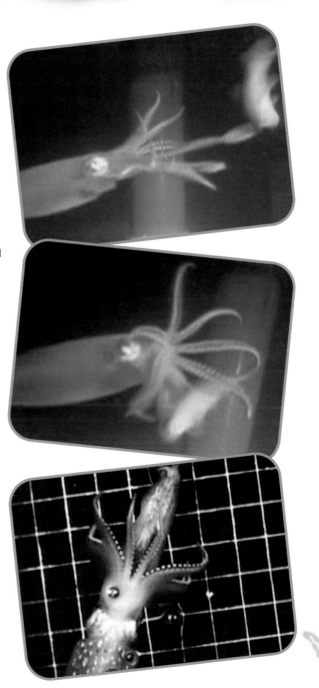

35

2

The **Elephant**

In any schnoz competition there's no doubt that the elephant would win by a nose. And it's this extremely versatile extremity that leads the elephant into number two in the body part countdown.

This is no ordinary nose. It's actually a complex array of over 100,000 muscle units that have the combined strength to lift enormous loads. And yet the trunk is capable of movements so delicate that it can be used to paint masterpieces. Perhaps that's why some humans feel their noses are a little inadequate.

Every year over 150,000 Americans have their noses surgically enhanced. But surgery isn't always the answer, as top Beverly Hills plastic surgeon Dr. Paul Nassif explains:

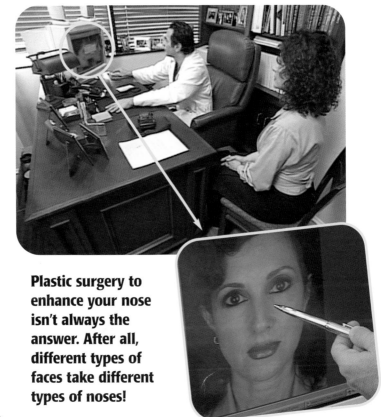

Plastic surgery to enhance your nose isn't always the answer. After all, different types of faces take different types of noses!

Some patients will come and ask me, can you give me a nose like Meg Ryan? Can you give me a chin like Ben Affleck? But in most of those scenarios we can't do that. If we did take those features and put them on a completely different person, it might not work with the face. We have to really do what fits the face.

37

The nose of the elephant fits its face so well that it would never opt for surgical modification. How could you hope to improve on a body part that's as useful as the human hand?

The two fingerlike projections on the tip of the trunk are so nimble they can pick leaves from tall trees and pluck grass from the ground. This is a hand that smells. Some scientists claim that the elephant's sense of smell is one of the most acute of any mammal. It's said to be able to smell water nearly twelve miles away!

The fingerlike projections on the tip of the elephant's trunk are so nimble that some elephants can even paint!

This family of elephants sure is thirsty. They drink with their long trunks, which help them take great big gulps of water.

And when an elephant gets to the water hole, there are plenty of other uses for that incredible trunk. It can suck up about a gallon of water at a time, and then become either a shower nozzle or a drinking fountain.

It's not easy learning to use such an extreme nose. Jumbo junior will need several years to completely master its trunk and all its uses. But not even the elephant's extraordinary appendage is the most extreme body part in the countdown.

1

The Giraffe

We've seen the nine contenders. They're the best of the best. Only one animal is a more extreme collection of body parts. The most extreme body parts on the planet can be found on the plains of Africa. Striding into number one in the countdown is the skyscraper of the animal kingdom—the giraffe. The word "giraffe" comes from an Arabic word meaning "the tallest of all," and it's easy to see why.

The world's tallest animal has some seriously stretched body parts. Its legs alone are taller than the average man. Its neck is even longer, giving it a great view of any potential predators in the grassland. Having a head for heights also means that the giraffe can feed in places no other grazers can reach. Even its tongue is enormous, plucking branches like a miniature elephant's trunk covered in sticky saliva. The giraffe has such a bizarre collection of body parts that it's difficult to get an idea of just how tall it really is.

Tall legs, a long neck, and a very sticky tongue help giraffes feed on the tasty leaves of even the tallest trees.

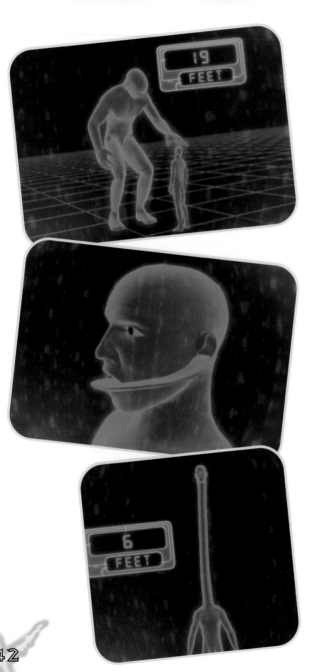

Imagine meeting a human as tall as a giraffe. He'd be three times taller than the average man—that's a towering nineteen feet high. Every basketball coach in the world would love a player tall enough to lean on the backboard!

If we were built like giraffes, we'd have another neat trick. With a twenty-inch tongue we'd have no trouble cleaning our ears.

The giraffe's most extreme body part is its neck. Incredibly, they have only seven bones in their neck—just like humans. But their bones are much, much bigger. Imagine if your neck bones were ten inches long! Then you'd really be able to stick your neck out!

A man built like a giraffe would stand nineteen feet tall, could clean his ears with his tongue, and would have a ridiculously long neck.

Drinking can be very tricky for the giraffe. You try bending over for a drink of water while walking on stilts!

But living with a six-foot-long neck isn't easy. If you think negotiating the muddy fringes of a water hole on stilts is tricky, just think of the difficulties involved in getting a drink of water!

Because it has to pump blood all the way up that extreme neck, the giraffe's heart weighs a massive 24 pounds. That's 35 times heavier than a human heart, and it pumps 6 times the amount of blood each minute. If it weren't for special valves in the neck controlling the flow of blood, the giraffe would pass out from the rush of blood hitting the brain every time it bent over for a drink!

There are some humans that find long necks so attractive that they've become known as the "giraffe people." In Southeast Asia, women from the Kareni ethnic minority elongate their necks by fitting them with brass rings.

The rings increase the distance from ear lobe to collar bone up to ten inches—that's more than double the average length of the human neck. The rings work by pushing down on the rib cage. This makes the body shorter so it looks like the women have longer necks. The only danger is if they're removed, because the neck muscles are so weak they can no longer support the head.

Known as the "giraffe people," Kareni women in Southeast Asia wear brass rings that stretch out their necks.

These male giraffes use their strong necks like sledgehammers as they fight over a female.

For the giraffe, a strong long neck is also the key to winning a mate. That's because romantic giraffes give a whole new meaning to "necking." Bull giraffes use their extreme necks as weapons. Each sledgehammer blow drives the head's two bony horns into the side of an opponent. It's no surprise that such an extreme collection of body parts should find such an unusual way to settle disputes. That's why there can be no arguing that when it comes to body parts, the giraffe really is The Most Extreme.

For More Information

Melissa Cole, *Elephants*. San Diego: Blackbirch Press, 2003.

Sam Dollar, *Anteaters*. Chicago: Raintree, 2001.

Sandra Donovan, *A Fox in Its Den*. Minneapolis, MN: Lake Street, 2003.

Richard Ellis, *The Search for the Giant Squid*. Minneapolis, MN: Sagebrush Education Resources, 1999.

Liza Jacobs, *Fiddler Crabs*. San Diego: Blackbirch Press, 2003.

Sandra Markle, *Outside and Inside Giant Squid*. New York: Walker, 2003.

Brad Matsen, *The Incredible Hunt for the Giant Squid*. Berkeley Heights, NJ: Enslow, 2003.

Barbara Keevil Parker, *Giraffes*. Minneapolis, MN: Lerner, 2003.

Ian Redmond, *Elephant*. New York: DK, 2000.

John Bonnett Wexo, *Amble Through the Expansive Grasslands of Giraffes*. Poway, CA: Wildlife Education, 2003.

Glossary

archaeology: the study of the activities of ancient humans

canid: the family of animals that includes foxes, wolves, and dogs

canine: conical, pointed tooth

dentition: teeth

formic acid: a poisonous acid found in ants

mucus: a slippery secretion

primate: the family of mammals that includes humans, apes, and monkeys

retina: the membrane that lines the eye

Index